I0639429

Sidney Coupland

Personal Appearances in Health And Disease

Sidney Coupland

Personal Appearances in Health And Disease

ISBN/EAN: 9783744717632

Printed in Europe, USA, Canada, Australia, Japan

Cover: Foto ©berggeist007 / pixelio.de

More available books at **www.hansebooks.com**

HEALTH PRIMERS.

No. 5.

HEALTH PRIMERS.

EDITORS'.

J. LANGDON DOWN, M. D., F. R. C. P.
HENRY POWER, M. B., F. R. C. S.
J. MORTIMER-GRANVILLE, M. D.
JOHN TWEEDY, F. R. C. S.

CONTRIBUTORS TO THE SERIES.

G. W. BALFOUR, M. D. St. And., F. R. C. P. Edin.
J. CRICHTON-BROWNE, M. D. Edin., F. R. S. Edin.
SIDNEY COUPLAND, M. D. Lond., M. R. C. P.
JOHN CURNOW, M. D. Lond., F. R. C. P.
J. LANGDON DOWN, M. D. Lond., F. R. C. P.
TILBURY FOX, M. D. Lond., F. R. C. P.
J. MORTIMER-GRANVILLE, M. D. St. And., F. G. S., F. S. S.
W. S. GREENFIELD, M. D. Lond., M. R. C. P.
C. W. HEATON, F. C. S., F. I. C.
HARRY LEACH, M. R. C. P.
G. V. POORE, M. D. Lond., F. R. C. P.
HENRY POWER, M. B. Lond., F. R. C. S.
W. L. PURVES, M. D. Edin., M. R. C. S.
J. NETTEN RADCLIFFE, Ex-Pres. Epidl Soc., etc.
C. H. RALFE, M. A., M. D. Cantab., F. R. C. P.
S. RINGER, M. D. Lond., F. R. C. P.
JOHN TWEEDY, F. R. C. S.
JOHN WILLIAMS, M. D. Lond., M. R. C. P.

HEALTH PRIMERS.

PERSONAL APPEARANCES

IN

HEALTH AND DISEASE.

BY

SIDNEY COUPLAND, M. D.

NEW YORK:

D. APPLETON AND COMPANY,

549 & 551 BROADWAY.

1879.

CONTENTS.

———◦◦◦———

PERSONAL APPEARANCES

IN HEALTH AND DISEASE.

INTRODUCTION.

IF a person wholly unacquainted with the structure of the body or with any of its functions could be confronted with a number of individuals, some of whom are what we call healthy and others what we call unhealthy, he would have very little difficulty in discriminating the one group from the other. The unhealthy ones might none of them be suffering from any grave disease, they might even be pursuing their ordinary avocations, and yet without putting a single question to them this unskilled, and possibly not very discerning, individual would have but little hesitation in making the broad distinction. He could not tell why he arrived at that conclusion, he might only say that these did not "look so well" as those; yet he would have gone through the process of picturing to himself what a healthy man should be, and would contrast his ideal with the forms before him. It is very likely indeed that here and there he might make a mistake, for being an undiscerning man he might be deceived by the appearance of health which some diseases give, or by not knowing the limits to which a body performing all its functions well enough to be considered healthy may exhibit a leanness to which he would feel inclined to apply the term of illness. Still, with some few exceptions, he would be in the main right. If he were asked to push his conclusions further, and to point out among the unhealthy ones those whom he deemed most and those least ill, and try and construct a scale of ill-health from the frames before him, it is likely that he would go very wide of the mark indeed.

Now it is the object of this little book to try and explain as briefly as possible how and why variations that are so plain on the surface can be taken as indices of disorder within, to give the reasons for form-changes which occur within the limits of health, and for those which mark the departure from those boundaries. It cannot be denied that this is a subject of very great importance; but it is beset with difficulties on all sides, difficulties such as those which the mere definitions of the terms "health" and "non-health" imply.

One great difficulty stares us in the face at the outset, and it is this: although built up on a definite plan, when viewed from the standpoint of the morphologist alone, the individual variations in the form of the body, slight though they be, are yet so numerous as to dispel once and for all any notion that there is an ideal of human form which can be described in so many words. Of course there are some who, "framed in the prodigality of nature," seem to approach to that ideal which we associate with the beautiful, just as there are others so ill-fashioned, "cheated of feature by dissembling nature, deformed, unfinished," as to excite in the beholder feelings of quite an opposite nature. And yet both may equally enjoy all that we know as health, and indeed the latter may, like the mis-shapen Richard, be endowed with a mental and a bodily vigour far surpassing the former.

Our subject then being "Personal Appearances" in their widest sense, there is no need to dwell upon individual variations; but dealing with the matter broadly, we shall have to ascertain what it is that contributes to the form, size, and colour of the body, and what is the significance of departures from the normal in these re-spects. The subject is a wide one. We can but touch upon its threshold.

CHAPTER I.

THE FORM AND SIZE OF THE HUMAN BODY.

THOSE animals which come nearest to man in physical conformation have at all times been objects of interest and of study. Their habits of life, no less than their bodily structure, have been investigated so far as opportunities have occurred, and even their mental attributes weighed in the balance with those shown to be possessed by the lowest of mankind, and not always to the advantage of the latter. In outward conformation the tail-less anthropoid ape bears striking resemblance to his human fellow-creature, and even in details of anatomical structure the differences are so slight as hardly to amount to generic distinction zootomically speaking. The precise points at which the anatomy of man touches that of the lower animals are very numerous. The grosser and more obvious differences are those which relate to the hairy covering of the body in apes as contrasted with man's comparative bareness, the relative proportions of the limbs to the trunk, the comparative size of the skull or brain-case, to that of the face, and, lastly, the erect stature of man and the consequent modifications in and development of the lower limbs. The following table, from Professor Huxley,* shows at a glance the difference between the length of the bony skeleton of the extremities compared with that of the spine. The spinal column is taken as 100 :—

	Male Bosjesman.	Female Bosjesman.	Gorilla.	Chimpanzee.	Orang.
Arm . .	78	80	115	96	122
Leg . .	110	117	96	90	88
Hand . .	26	26	36	43	48
Foot . .	32	35	41	39	52

It will be seen that the upper limb in these three anthropoid apes is relatively longer than the lower, whereas

* 'Man's Place in Nature.' London, 1860, p. 71.

in the Bosjesman the reverse holds. In fact, in these apes the foot is used largely as a prehensile organ, and the usual mode of progression is, not in the erect posture, but in the semi-erect, or even on all four limbs.

The erect posture of man largely modifies his form by calling into play certain muscles which become more developed in him than in animals. The muscles of the spine, and all those in front of and behind the thigh which are employed in moving the lower limbs in progression are thus markedly developed, and their development forms an integral part of the contour of the body. But mere muscularity alone does not suffice to conceal the angularities of the bony framework to which the muscles are attached, or at any rate does not lead to the appearance of the fine curves and rounded outlines of the human form. All the deficiencies in this respect are supplied by the soft fatty tissue which underlies the skin filling up spaces left between muscles, and concealing bony prominences. There are thus three factors contributing to the form of the body—bone, muscle, and fat; and a more detailed description of each of these and of their arrangement is necessary before treating of the modifications they undergo, and the consequent changes in the general form of the body.

The regions into which the body is divisible are, head, trunk, and extremities. The head, containing the brain and bearing the sensory organs (which, with the jaws and mouth, go to form the face), is largely bony, the soft parts being very slightly developed over the head proper, but to a greater degree over the face.   The trunk, divisible into neck, chest, abdomen, and pelvis, has its form laid down in bone, which in the neck and abdomen is represented in the spine only, in the chest by the ribs, which attached to the spine behind and the breast-bone in front form the walls of the thoracic cage in which are contained the lungs and heart, with the air passages and blood-vessels connected with them, as well as the gullet passing on its way to the stomach.   The chest cavity is separated from that of the abdomen by a muscular septum, the diaphragm ; and contained in the abdomen are the organs of digestion, assimilation, urinary excretion, and others. The lateral and front walls of the abdomen are soft and pliable, being wholly composed of skin, subcutaneous fat, and muscular layers.   Lastly, the pelvis is flanked on each side by the greatly expanded haunch-bones to which the bones of the lower extremity are attached.   The four limbs, subdivided into segments and terminating in the hands and feet, are connected to the trunk by muscles and ligaments binding the joints.

The Skeleton.—The bony skeleton (Fig. 1*) or frame-

* For this and for many of the other illustrations we are indebted to Mr. H. E. Cree.

work consists of two hundred separate bones, eighty-three of which are in pairs, and those that are single are symmetrically shaped and could be divided into two equal and similar halves. The bones forming the head or skull are arranged into those which form the cranium proper and those which enter into the face. The cranium or brain-case rests upon the spinal column by articular surfaces. Its roof and sides are formed by several flattened bones, which in the adult are firmly interlocked one with another. Although expanded above, at the base of the skull, the bones are compact enough, and, except for the numerous perforations to allow of the passage of nerves from the brain and of blood-vessels, and a large aperture below

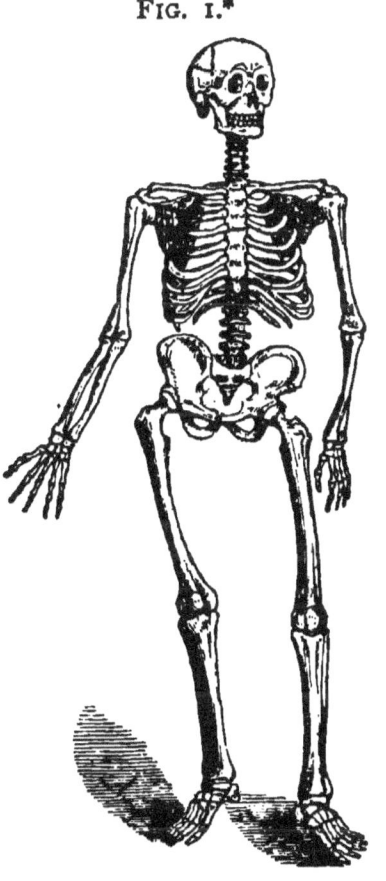

FIG. 1.*

* Adult male human skeleton.

for the upper part of the spinal cord, which is continuous with the brain,—the brain-case is a closed box. In front of the cranium is the face, the broad features of which are laid down in the numerous bones entering into its composition; bones to complete the eye-sockets, to form the foundation of the nose, to give the prominence to the cheek, and to form the upper jaw and palate; whilst articulated to the temporal bone is the lower jaw, which in the adult is squarish in front, and is acted on by powerful muscles for its closure (Fig. 2).

FIG. 2.*

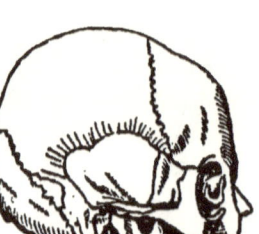

The spinal or vertebral column, or backbone, consists of a number of bones called "vertebræ," joined together, and capable of a certain amount of movement, which gives flexibility to the whole column. In the neck these are seven in number, and here their movement is greatest. In the dorsal region there are twelve, each of them bearing a rib on each side; then come five larger lumbar vertebræ, and four more fused together into one bone, the sacrum, and, lastly, three rudimentary bones, the relic of the tail. The cervical, dorsal, lumbar, and sacral consist of an anterior solid part or body, and an

* Adult male human cranium.

arch of bone behind, with a projection in the middle line, so that when united there is a channel through the whole of them, which lodges the spinal cord, passing from the brain. The shape of each bone varies slightly from its fellow, and very much according to the region in which it is placed, so that the width of the column is greatest in the lumbar region, where the bones are larger and deeper. United by fibrous bands or ligaments, which yet allow of their movement, they form the whole column, which is not straight, but is curved slightly forwards in the neck, backwards in the back, and again forwards in the lumbar region.

There are twelve ribs on each side of the chest; each of these articulates with the spinal column behind, and is capable of an upward movement at the joint. The first rib is flattened from above down, and is comparatively short. It passes from the first dorsal vertebra to join the upper part of the breast-bone or sternum, a flattened, sword-shaped bone which completes the chest in front. The other ribs are flattened chiefly from side to side ; six of them are in connection with the sternum in front by means of cartilages which are about two inches in length, and in shape exactly like the bony rib. The next three ribs are attached by their cartilages, not directly to the sternum, but to one another; whilst the eleventh and twelfth ribs—the latter shorter than the former—are quite free from anterior attachments, and are hence called "floating ribs."

The bones of the upper extremity consist in the large single arm-bone or humerus, the upper extremity of which is rounded and fits into a socket in the blade-bone or scapula—a flattened bone, which rests over the upper ribs on the back, and can move over them for some distance; a curved bone forming a prominent feature at the top of the chest—the "collar-bone" passes from the scapula to the sternum. The arm-bone is joined at the elbow by a hinge joint with the bones of the forearm, or rather with the ulna, which alone enters into the elbow-joint. The ulna is thicker at its upper extremity, where it forms the prominence of the elbow, than at its lower; its fellow, the radius, on the outer or thumb side of the forearm, articulating with the ulna above, and capable of rotation, alone bears the wrist. Then comes the wrist-joint and a series of small bones, eight in all, forming the wrist; to these succeed five long bones forming the framework of the hand, and upon them are articulated the finger-bones or phalanges, of which there are in all fifteen, three to each finger and two to the thumb. Each successive lower segment is smaller than the upper, the terminal phalanx being expanded at the extremity where the nail is borne. The length of the fingers depends upon the development of these bones, and the long, tapering fingers, and small hand of the high-born lady owe their shape to the delicately-formed skeleton ; just as the broad and short hand and clumsy-looking fingers of the rustic are due to his coarser and broader bones. Notice that the second, or middle finger, is the longest; that as a rule the third, or ring-finger, is longer than the first, but not always, and that the fourth is shortest of all.

The bones of the lower extremity resemble somewhat those of the upper. The great hip bones, which are united to the sacrum behind, and to one another in front, represent the shoulder-blades, and each bears a socket for the reception of the upper end of the thigh bone. This single bone, called the femur, is succeeded at the knee joint by two bones, one large and massive, and prismatic in shape—the tibia—which comes close under the skin in front and forms the shin. The other is slender, and passes on the outer side of the tibia, without entering into the knee joint, and flanks the outer side of the ankle. Then the bones of the foot, seven in number, interlocked into one another, one forming the heel, and bearing on its upper surface the square-shaped bone which articulates with the tibia to form the ankle joint. Then in front of this a boat-shaped bone, and then a series of four small wedge-shaped bones, which in their turn articulate with five long bones forming the front part of the foot, and these again with the phalanges

2

of the toes, stunted and small in comparison with those of the fingers. One other bone should be mentioned in the lower limb—the knee-cap, which is visible under the skin, and in front of the knee joint.

In the disease of early life called Rickets, when the skeleton is unnaturally soft,—the yielding bones of the thigh and leg give way under the weight of the child when

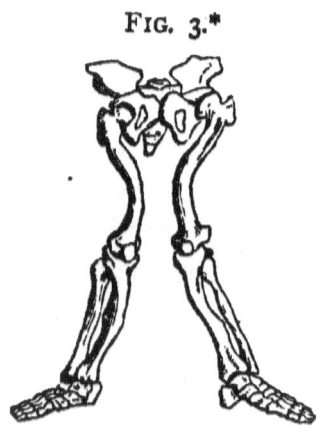

FIG. 3.*

it attempts to walk, and extreme and permanent deformity may result unless the child is kept off its legs. Thus the thigh bone curves forwards, and outwards, and the leg bones, especially the larger one, bend forwards, and bowed legs and "knock-knees" (see Fig. 3) result. Happily the soft and yielding nature of the bones allows of appliances early had recourse to effecting the straightening of the limb; but unless such measures be taken, and if the child continues to walk, the deformity will be increased, and as the bones grow firm will be borne by it throughout life.

Such in merest outline is the bony framework of the skeleton. It is completed by the union of the bones in

* Ricketty deformity of bones of lower limbs.

joints, which are of various forms and construction in accordance with the movements they have to subserve. Essentially, a joint consists in the two ends of the bones which enter into its formation, and which are covered by a layer of cartilage, a sac, secreting fluid which bathes the surfaces of the cartilages, and in fibrous bands or ligaments protecting these structures, and serving to keep the bones in apposition and to limit their movements. The expanded ends of the bones entering into the joints add to the diversity of the form of the body, and alteration in these joints themselves gives rise to conspicuous deformities.

The Muscles.—The skeleton, bound together and compacted by the ligamentous structures, is clothed with muscles, which are structures of flesh endowed with the power of contracting, and by their contraction pulling on the bones to which they are attached, and thus subserving the motion of the body. According to their development, so do these muscles play a large part in producing the outlines of the body, and their bold configuration is well reproduced in the sculptures of antiquity. Those muscles which are most wanted for the support of life, and for locomotion, are the most developed, e.g. those of the lower limbs.

Besides contributing largely to the mere contour of the frame, the muscles have an important part to play in maintaining the body in the erect position. The whole framework of the body is, as we have seen, composed of a large number of separate parts. Each of these parts has relations with others, and in many regions one can be moved on the other by means of muscular action.

There is moreover a state of equilibration between the muscles performing opposite kinds of movement; and so long as the muscles are in a state of tonicity, and so long as the one set bears its due amount of work as compared with the other set, so long is this equilibration maintained.

This may be readily illustrated by the part played by the muscles placed before and behind the spine, in maintaining the erect posture of the body. The position is kept up without effort, without even consciousness, by the healthy man whose muscles are well balanced and in good "tone." It may be, however, that the same man after a long day's work over a desk, in an. ill-ventilated city office, no longer presents that supreme unconsciousness of his muscles and their action, and the stoop of his shoulders and bent head demonstrate to others that the balance is no longer kept, that the tonicity of the morning has passed off, and the wearied muscles are no longer on the watch. And so it is when in sleep the muscles are relaxed and gravity asserts its force, so that the head falls forward by its own weight, no longer restrained by the passive counteraction of its "extensor" muscles. It may be mentioned in passing, that so little is there any effort required to maintain the body erect, that it is a sign rather of weakness than strength in anyone who exercises an effort to do this. This may seem paradoxical, but it is nevertheless the case; and he who walks "bolt upright" with his chin in the air and his back as rigid as a plank, is often not a strong, but a weak man.

In consequence of this it happens that various deformities arise from the equilibrium between opposing muscles being destroyed from weakness or palsy of the one set, or by the over-action of the other set. For instance "wry-neck," which consists in the head being permanently fixed in a position inclined to one shoulder, or its continual spasmodic movement in that one direction, is due to the contraction, or over-action of the muscle causing that movement not being balanced by its opponent; or if the muscles of expression, those of the face, be palsied on one side, the mouth appears drawn up and out to the opposite side, an appearance exaggerated by calling the muscles into action as in laughing; or, again, in "squinting," we have an example of weakness or palsy of one or more of the six small muscles moving the eye-ball preventing the movement of the eye in their direction of action, so that the two eyes no longer move in concert, and lose their parallel action. Change of form is also obvious enough from wasting of particular groups of muscles. We may only instance the flattening of the back of the forearm in lead-palsy, where the muscles which bend the hand backwards are especially attacked, so that the hand falls and cannot be raised—the condition known as " wrist-drop."

Then again, weakening of the muscles of the back (as produced by enforced sitting on benches without support to the back) is one cause of what is called lateral curvature of the spine, where the curve to one side which the spine has in the loins is greatly increased, so much so, that the shoulder on the side to which the curve is directed is lower than its fellow; and a curve in the opposite direction, causing projection of the blade-bone, forms in the dorsal region.

Club-foot is in a large majority of cases due to muscular defect—a defect which may be present to a certain degree at birth, or which may arise from an unnoticed paralysis in early childhood. The most common form of this is, that where the heel is raised, and the subject walks not on the sole but on the toes, or more frequently on the side of the foot. The strong tendon which passes from the calf muscle to the heel is in such cases contracted, the muscles themselves being wasted from their long enforced inactivity, so that the leg appears small and shrivelled.

Nor simply from bearing weight do ricketty bones give way. The muscles that are attached to them act as powerfully in producing their distortion, as may be seen in the thigh bones (Fig. 3), in the collar bones, and in those of the upper limbs, the powerful muscle which passes from the shoulder blade to be attached to the humerus on its outer surface often causing considerable distortion of that bone.

Such are some of the changes of form which muscles give rise to; those depending upon great muscular development are too obvious to be detailed, the fact need only be stated that the more work a muscle is called on to perform the larger it grows, and as it grows, it must proportionately modify the form of the body.

The Fatty Layer.—Lastly, beneath the skin is the layer of fatty tissue which is in very variable amount in different sites, being for instance wanting in the scalp, which is in contact with the skull, but abundant in the face of a well-nourished individual, concealing the prominent cheek-bones. Fat also occurs in the interior of the body around organs, but this really does not concern us here. However, it may be mentioned that the eyeball rests on a cushion of fat; and the sunken eye of a wasted person is due to diminution in this cushion. In addition to this, the natural fulness of the skin is further due to the presence throughout it, and the textures underlying it, of blood-vessels full of blood ; this has a great influence on the shape—as seen by the shrivelled aspect of a part which is deprived of blood from any cause, e.g. cold, which causes contraction of the small arteries, and lessens the supply of blood to the part.

The general form is dependent on the due relation of all these parts, and modification in either will lead to proportionate change in shape of the body.   The shape of

Fig. 4.*

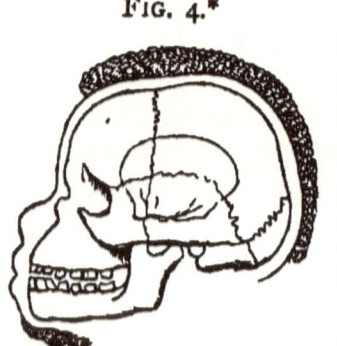

the head, as a whole, is determined by the form of the skull, as is well exemplified in the negro (Fig. 4), where the protuberant and prominent lips are attached to the projecting upper jaw-bones — the prognathous form of skull.

It may be observed that there is a perfect bilateral symmetry in the whole body—a symmetry which is not only apparent on the exterior, but obtains in the skeleton, and to a great extent in the disposition of the internal organs.

We may now rapidly enumerate the leading changes in form undergone by the body at different periods of life.

* Skull of negro, showing prognathism.

In *infancy* the head is large in proportion to the chest, and the bones of the cranium are not fully ossified, but are separated from one another by membrane, which, between the frontal and parietal bones, and between the latter and the occipital, leaves a considerable interval. Then the face is small in proportion to the cranium, and the lower jaw, instead of having an angle at its posterior extremity, which in the adult is almost a right angle, is nearly horizontal (Fig. 5). The skeleton is incompletely ossified, the muscles small and ill-developed, the contour of the body being chiefly made up by the excessive fatty layer which underlies the skin.

FIG. 5.*

The neck is short; the chest prominent, narrow, and short; and the abdomen large and capacious, whilst the spine is nearly straight, and the lower limbs proportionately short. In *youth* the chief change lies in the greater development of the lower limbs, the greater proportional increase in the size of the chest, and corresponding diminution of the head and abdomen. The spine now forms a double curve. In *adult* life, when all the skeleton is firmly ossified, the jaw is square, its angle almost a right angle (Fig. 2), the chest fully expanded, the lower limbs well developed. In *old age* the

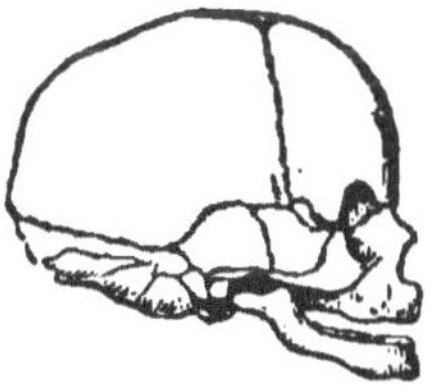

* Skull of infant.

stature is less, the neck broad, and shoulders rounded; the bones are more fragile, the rib cartilages fixed and hardened; the lower jaw, from loss of teeth, reverts somewhat to the infantile form (Fig. 6), whilst the chin appears

FIG. 6.*

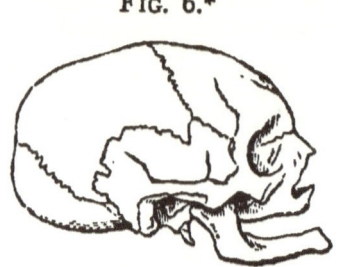

more prominent and approximates to the sharpened projecting nose; and the skin hangs loosely, and in folds and wrinkles, in proportion as the fatty layer has decreased and the muscles have diminished.

Of the differences in the two sexes suffice it to note that the average stature of women is less than that of men; that in the former there is a greater tendency to the growth of the subcutaneous fat, and that the skeleton of the pelvic region is modified so that the hips are wider in the female than the male.

CHAPTER II.

CHANGES IN THE FATTY LAYER.

HAVING thus as it were propounded the bases upon which the form of the body depends, let us consider

* Skull of aged person. It is unusually long from before backwards in the above figure.

the variations which this form undergoes. Some of these are irremediable in the individual, stamped upon him by inheritance ; others are the effects of declared disease deranging the economy ; others again arise from the perverse practices of mankind in the matter of clothing, or by the equally perverse pursuit of certain modes of life and labour which cramp and distort the frame, disabling the individual for the comfort of the community. Such and many more comprise a long catalogue of effects, some of which are compatible with good bodily vigour, whilst others can only be regarded as warning signs, if not actual indications of defective health.

Perhaps the most convenient way of dealing with the matter will be to discuss seriatim the effects which arise from changes occurring in each part of the framework, and show how deficiency here or redundancy there may be brought about, and to indicate what it teaches.

We have then first to treat of changes in the Fatty Layer.

It is impossible to say where leanness ends and fatness begins, just as it is impossible to mark the dividing-point between any other two opposed states. But all the world knows what is called a "lean" figure, just as it can point to a corpulent one, only between the two there is of course any amount of gradation. In the absence of any given standard of proportion between the height of the body and its girth, we need not attempt a strict definition of either one or the other, but must be contented with the general statement that leanness depends on a deficiency in the soft parts, and corpulence upon an excess of the fatty layer especially. Now each of these conditions, even to a very extreme degree, is compatible with health. Many, indeed, if not born lean, yet acquire leanness because they inherit the disposition to it; these are the wiry muscular men, and the angular women. To have a "lean and hungry look" does not by any means necessarily portend that the possessor "thinks too much." It may simply mean that he has inherited the type from his parents, and were he now submitted to the close confinement and rich fat-making diet of a Strasburg goose it is doubtful if the wanting adiposity would ever come to round the angles of his frame. It must then always be borne in mind that deficiency in fat is not a sign of disease, unless the individual has been previously well nourished, and although there is no doubt that much is claimed for the influence of heredity which it can hardly bear, yet this goes for a great deal in the production of a spare habit of body. Many other factors, however, help in its production; continued muscular activity is a well-known counteracting influence against the over production of fat, for the muscles in their work require nutriment, taking it from the food, which unless disposed of in this way would go to add to the general store of fat. And accompanying with such exertion there are the quickened circulation and increased activity of respiration, by which the blood is more rapidly exhausted and requires more frequent renewal than when the body is at rest. When to this is added a spare but sufficient diet, and especially an avoidance of malt and alcoholic drinks, the induction of a lean habit is more and more favoured. There can be no doubt also that equally with muscular activity, continued mental exercise will hinder fat production, for although it would be an unfounded aspersion, frequently disproved in fact, that fat people have sluggish faculties, yet there can be no question that severe mental application tends to a lean habit, and that the words put into Cæsar's mouth by Shakespeare in speaking of Cassius have some measure of truth in them.

It may be pointed out here that as a rule the inhabitants of tropical climates are leaner than those of temperate or frigid zones, and the probability is that the former make less flesh. However, there are many exceptions to the rule; for the indolent life of the first named, dwelling in a land " where it is always afternoon" and where nature in her luxuriance is always at hand to supply his physical needs, would favour rather the tendency to fat production when contrasted with the more active work of the latter.

One more point about lean people and we pass them

by for the more interesting group. It is notorious that persons of spare habit are not only capable of resisting more fatigue than others, but they often show far more resistance to the inroads of disease. This is especially the case with acute diseases, e.g., fevers. As a rule there is more vitality, more ability to struggle with the violent disorders of nutrition in such diseases among spare people than among the stout, so that a lank ungainly habit has its compensation,—it is more often tenacious of existence than the apparently but not really more robust frame.

When we turn to the opposite condition to leanness we find ourselves trenching upon the borders of disease, for that beyond a certain extent corpulence is non-natural may be admitted, and that it is dependent on defects in the working of the animal mechanism is highly probable. In infancy, fat production is paramount ; the food of the infant is rich in materials for its formation, and although growth is proceeding with marvellous rapidity, yet a thin babe is an abnormality, a witness to improper feeding so common among all classes. The inert, almost vegetable, existence of the suckling infant favours this storage of fat, and its plump cheeks, its chubby extremities marked by deep furrows at their flexures, are indications of this storage. But as has been pointed out, even in this early period of existence the production of fat may exceed the normal, and trangress the bounds of health ; and then all the forming causes tell with tenfold vigour.

But as growth proceeds, and activity takes the place of inertness, the tendency to the production of fat is lessened ; and hence it is that a fat child or youth is rightly regarded as one whose nutrition is at fault, or whose diet is of a kind which should not be encouraged. The full exercise of the body, during its period of growth, is incompatible with the production of fat, and strongly counteracts what tendency there may be to its formation. As life advances, and the period of growth is passed, and maturity is yielding before the advance of old age, then the lessened call upon the expenditure of the body favours the accumulation of fat, and we reach that period of the seven ages when corpulence is regarded almost with reverence, and even desired as adding dignity and importance.

Of the causes of corpulence that which is predominant is that to which we have already referred with regard to leanness. The tendency to obesity is as much under the influence of heredity as the tendency to spareness. It " runs in families," and it can as little be avoided as a man can get rid of his features or the colour of his hair, or tendency to early baldness, or any of the thousand and one proclivities which he derives from his parents. Vain indeed is the effort to reduce his form to greater comeliness. No amount of abstinence in diet, of regulated or even excessive exercise will, except at the expense of health, rid him of his proclivity, although care may keep it within limits.

There are, however, many subordinate and favouring causes to corpulence. The first to be mentioned is that of sex. Undoubtedly women are more prone to this production than their hardier brothers. Without seeking for an explanation of this fact, it is better to accept it as a fact, true from the beginning, and parallel with the facts known as to the average stature of the sexes, that women have more fat than men.

So again, the time of life, as we have already indicated, is a predisposing cause to corpulence.

But there are other causes, not indeed often if ever sufficient of themselves to increase the amount of the fat in the body, fat which, be it remembered, is not limited simply to the exterior as we are now considering it, but abundant also around certain organs and in certain tissues in the interior of the body. These are, first and foremost, the continued consumption of rich oleaginous, starchy or saccharine food, and drinks, such as malt liquors and alcohol. Who is not acquainted with the typical drayman, whose portly figure matches well with the rotundity of the barrels on his dray, and the plump sturdy horses between the shafts? Physiologists have calculated to a nicety the number of ounces of nitrogenous, carbonaceous, and fluid, material that is necessary to be the daily food of man, to supply his normal waste, to give the amount of fuel sufficient to keep the human mechanism in due working order. All over and above this is waste, much thrown off, much also stored up as fat to be called for if occasion ever require it. Hence regulating diet to a nicety, by cutting off excess of those foods which most go to "make fat," and such liquors that do the same, was the gist of that method aiming at the cure of corpulence propounded to Mr. Banting by the late Mr. Harvey, and quoted now in learned medical treatises as the system of "Bantingism."

Another cause, and one which the very possession of a corpulent habit only serves to perpetuate and extend, is indolence—indolence of body and indolence of mind. To take things easily, "to laugh and grow fat," to do nothing with undue expenditure of force or thought, this is one of the main causes of fat production. And many people, already corpulent, cannot help being lazy. It is a greater effort for them to pursue even the ordinary avocations of life, how much more its more arduous and less necessary exertions! Lack of energy is then both cause and consequence, and as the fat grows the muscle wastes; and to "make flesh" is a phrase which often should more properly be used in the opposite sense.

3

Other favouring conditions for the development of fat are equability of temperament, the cold, impassive "phlegmatic" rather than the excitable and nervous; and an impoverished condition of blood whereby its functions are ill-performed, and that function of getting rid of waste material also.

Corpulence is an attribute which is by no means unmixed with the reverse of blessings. Its greatest evil is that its possessor when attacked by acute disease is less fitted to contend against it than his sparer neighbour. This may be in part because, as we have seen, habits unhealthy in themselves are so often at the foundation of the corpulent state. But even apart from this and from the difficulties the fat man has to contend with when laid low by illness, it must be generally admitted that his power of resistance to acute inflammatory disorder is but slight.

We come now to conditions that are abnormal, both in diminution and in excess of the subcutaneous fatty layer. Many diseases, nay most, are accompanied by "loss of flesh." In some affections this wasting takes place rapidly—e.g., in the acute febrile diseases; in others it is slowly progressive. It depends upon two factors, one excessive consumption of tissue, the other diminished capability of assimilation and of nutrition. In a few maladies both these factors are at work; with a chronic wasting disease, such as consumption, it is the latter only or almost solely. To deal first with the latter cause, viz., diminished assimilation of food, let us see the effect that is produced by starvation. It does not require a visit to a famine-stricken district to become acquainted with the repulsive details. Examples enough are unfortunately presented daily before our eyes, where from ignorance of the simplest rules of diet, infants are practically starved, although apparently highly fed. And it is in infants that the effects of deprivation of food, or of the administration of such food which cannot be assimilated, and therefore cannot subserve nutrition, is most plainly seen. The first tissue to "go," and that which wastes to its largest extent is the fatty layer beneath the skin. It has been calculated that as much as 93 per cent. weight of fat can be lost in the process of starvation—a degree of wasting which far exceeds that of any other tissue in the body. It is as if this material is standing in reserve, to be first utilised when either food is withheld, or when the processes of disintegration surpass those of growth.

The rapidity with which the transition from the fulness and rotundity of health passes into the more or less highly attenuated condition is exemplified in several affections, which, being of an acute character and accompanied by considerable discharges from the blood, e.g., cholera, drain the blood-vessels and take away fluid from them.

And these effects are shown more rapidly and readily, although perhaps carried to a less extent, in individuals who previously were well nourished or even corpulent. For in these the contrast is all the more striking than in those in whom from the slight amount of adipose tissue normally present there is already the appearance of wasting. Hence it is that the appearances of ill-health are best seen in those who previously had the greater appearance of robust health, and the sunken cheek and drawn face show early the inroads of disease.

In acute fevers there is a great call upon the stores of nutriment in the body, and much waste takes place which cannot be made up while the fever process lasts; because the functions of digestion and assimilation are perverted. Hence in a few days or weeks of fever the body emaciates to a marked degree. Nor is the wasting limited to the fatty layer. As in starvation this doubtless is the first to be used up, but then further call is made upon the other structures, and muscles grow thin and flabby too. The recovery of form after the fever has passed away, during the stage of convalescence, is chiefly by the reproduction of fat; for it requires a long period of careful exercise to restore the muscles to their previous natural condition. Indeed, it is frequent enough that an acute and wasting fever will cause such permanent alteration in the nutrition of the body as to lay the foundation for a condition of corpulence to which the subject was previously quite a stranger. The slower process of losing flesh that occurs in chronic wasting diseases affects the same parts, and depends either upon actual starvation, the disease involving the digestive organs, or upon some interference with the natural transformation of food into tissue, or the occurrence of prolonged discharges impoverishing the blood.

Mere modifications in quantity of the subcutaneous materials are not the only causes of change of form in the soft parts; abnormal products may accumulate in their tissues and lead to change of shape. This is seen, e.g., after an injury, when a swelling is produced by the escape of blood into the tissues; or when a part is inflamed. Then it becomes swollen by the escape not actually of blood but of some of the constituents of the blood, and may even go on to form a limited swelling full of matter, an abscess. Or tumours of all kinds may give rise to local derangements in shape. But there is a condition, in dropsy, which may effect a change in the whole body, or may be limited to one part. Here the serum of the blood which is unduly watery escapes into the tissues, and replaces the fat; and the skin becomes stretched and smooth in proportion to the amount of the dropsy. Seen to its greatest extent in the dependent lower limbs, where the fluid naturally gravitates, it may appear in all parts, as in the loose tissue under the eyelids. Many conditions give rise to this, some depending on local disturbances of the circulation, others on general blood conditions, and thus it may be of grave and serious significance.

Lastly, in the subjects of long-standing disease of the heart or of the chest, there occurs often what is called "clubbing" of the extremities. It is best seen in the fingers, the tips of which become broad and rounded, especially the soft palmar surface. But the same clubbing may be seen at the extremity of the nose. In all cases it is dependent on the slowing of the circulation which takes place in such diseases, so that there is an almost permanent condition of overfulness of the minute blood-vessels of these extremities.

CHAPTER III.

CHANGES IN THE BONY FRAMEWORK.

HAVING sufficiently alluded to changes in form produced by alterations in the muscular structures (see p. 17 and *seq.*), and in the fatty layer (see p. 24 and *seq.*), we have to consider now those due to alterations in the bones and in the joints. Changes which occur in the shape of the skeleton necessarily entail very marked deformity. Some of these are the result of previous muscular failure, others are due to affections of the joints, others to alterations in organs contained within or in contact with the bony structures, and others, again, to altered conditions of the bones themselves, arising primarily, and independently of any previous condition of contiguous parts. When a bone is broken there may or may not be considerable deformity; its occurrence depending on the nature and extent to which the injury has taken place, the position of the bone itself and its connection with muscles. In consequence of this, sometimes most careful scrutiny will fail to detect that a bone is broken, whilst at other times the whole limb may be bent and the injury be manifest at once. Thus a rib may be broken, but no external change in shape will reveal the fact to the observer. But the arm or thigh may be fractured, and the amount of deformity be striking. The more oblique the fracture is, the greater the resulting deformity; for then the muscles attached to the fragments will in their traction on each pull them the one over the other and so lead to shortening of the limb.

Sometimes when the two fragments of a broken bone do not grow together again naturally, or unite in a distorted manner, the limb remains permanently misshapen. But even where the union is natural and perfect, an irregularity in the shape of the bone itself will denote long after the spot where the junction has taken place.

The main cause of distortions in the shape of the bony

framework, and therefore of the whole body, is the disease of infancy called rickets, to which allusion has already been made. In this affection the whole skeleton, at its period of most active growth, when the soft carti-laginous and membranous structures which are ultimately to become firm bone, still form a large part of the framework, suffers from lack of proper material for the formation of bone, and grows exuberantly but yet not firmly, and the evil results that ensue are due in great measure to the unnatural softness of these structures combined with their great abundance. There is abundant material to form bone, but it is bad material; there is active growth, but it is a perverted activity.

Some of its effects have already been mentioned; others will be met with later on. One of these effects is a curvature of the spine, not to one side, as we saw resulted from muscular weakness, but a bending forwards, giving the appearance of the spine "growing out," as it is popularly termed. A sharper angle in this direction, leading to considerable deformity, is produced by actual disease of the bones of the spinal column. Such curva-tures are called "angular curvatures," distinguishing them from the "lateral" bendings before mentioned. "Knock-knees" and "bowed legs" are mostly due to rickets (see Fig. 3); their names sufficiently indicate the deformities they produce.

Any account of the changes of form undergone by the body would be incomplete, even in a general sketch like this, without some mention of those partial but frequently troublesome and permanent distortions which arise from affections of the joints. All the structures entering into the composition of a joint may at one time or another be so altered as to lead to deformity. Thus, the synovial sac (see p. 17) may become distended with fluid, and give the joint a swollen rounded appearance, concealing entirely the natural prominences of the extremities of the two bones that form it; or the fibrous ligaments surrounding the ends of the bones and keeping them together may be swollen and altered from inflammation, and lead to similar swelling and alteration of the other soft parts, so as completely to disguise the joint and render it more or less fixed. Or these same ligaments may be shortened and changed by disease so as to cause one bone to pass beyond the other, and produce what is called a dislocation, or the dislocation may take place from a violent injury, without any disease of the joint, and lead to marked deformity of the limb at the part of the displacement. Or the ends of the bones themselves may be enlarged and the seat of changes, which permanently cause an alteration in the shape.

As to simple dislocation from injury, it may be remarked that in some joints this cannot take place unless bone or ligament be broken, so intimate is the interlocking and union of the normal parts.   But on the other hand there are joints of which the structure admits of ready displacement.   When a joint is dislocated, this is shown by the appearance of the end of the displaced bone outside the socket, or separated by an interval from the extremity of the bone with which it was previously in contact.   There is then such a deformity as results from the disappearance of the rounded end of the bone at one place, and its appearance at another, a depression occurring where formerly there was a projection.

There are, however, affections of the joints which are common enough to be frequently observed, and concerning which a few words must here be spoken.   When from a twist or a strain the ligaments around a joint are injured, i.e., when a sprain, as it is called, takes place, there is always a certain amount of inflammation set up. This may be, and indeed if the sprain be a severe one, is, preceded by considerable bleeding into the soft parts around the joint, and a consequent bruise over the swollen joint is added to the other evidences of injury. The amount of force exerted on the joint, or the direction of the force has not been enough to cause a dislocation, but has sufficed to stretch or rupture ligaments and give rise to the inflammatory changes.   A joint may become swollen and inflamed from other forms of injury, as direct blows or wounds.

But the inflammations set up by diseases such as rheumatism and gout, are definite in their characters, are productive of great alteration in the form of the joint, and may lead to changes of à permanent or chronic character. Let us watch the course of an attack of rheumatic inflammation, say of the knee, in so far as it causes changes of form.* The pain which foretells and which indicates the seat of the affection is soon accompanied by perceptible swelling; there is a rounded fulness of the knee, and if the synovial sac, as often happens, has a large quantity of fluid secreted into it, it is specially distended, and bulges beneath and on each side of the kneecap. The surface is slightly reddened, a redness which fades on the least touch, under the pressure of the finger. Or the ankle also may be attacked, and the bony projections on each side of it, which are formed by the lower ends of the two bones of the leg, are concealed in the puffiness and smoothness of the inflamed tissues. And here may be mentioned the fact, not, however, without very many exceptions, that the larger or middle-sized joints, such as the knee, ankle, wrist, and elbow, are more liable to be inflamed and swollen in acute rheumatism than other joints. Still frequently enough the

---

* It must be distinctly understood that the purpose of this little book is not to give full details of the symptoms of any disease, except in so far as any change in the normal form is brought about by the disease.

smaller joints of the fingers, and less frequently those of
the toes, also suffer. In a short but variable time, a few
days or at most a few weeks, the swelling subsides; with
the subsidence of the inflammation, the joint regains its
natural shape and mobility. But the rheumatism may
become chronic, and then the swelling, although much
diminished in extent, will yet remain a very long time, or
may indeed never wholly disappear. And in this connection we might refer to an inveterate, nay an incurable
and painful malady, one which never loses its hold on
the sufferer, but advances as it were step by step from
joint to joint, cramping their movements, and finally entirely restraining them, and leading to an amount of distortion and alteration in and around them which entirely
alters their form as well as their function. Under its
popular cognomen of " rheumatic gout," or its scientific
name of chronic rheumatic arthritis, this affection is known
to many. Happily of not very frequent occurrence, its
inveteracy is its foremost feature, an inveteracy due to
the slow disorganising changes to which it leads. The
knuckles of the hand become rough and irregular, and the
small bones of the fingers are displaced from their natural
relations one to another. The nodular prominences
which occur over and above the normal outline of the
joint are due to the production of fresh bony substance
in the ligaments and from the bones around them, while

the articulating surfaces of the bones gradually lose their cartilaginous covering, and become slowly altered in shape. In consequence of the fixing of the joints, the muscles which acted on them are no longer called into play, and these waste, and lead to a shrinking in the bulk of the limb itself.

The changes in shape of the joints which gout produces are somewhat like those of rheumatism. (The joints affected are as a rule those of the great toe, ankle, fingers, but even knees, wrists, and elbows.) In the acute form of gout, if it were not for the particular joint implicated and collateral evidence, some difficulty might first exist in determining whether the joint swelling were due to rheumatism or gout. The skin over a gouty joint in the acute stage is glossy and shining, and when the disease becomes chronic the deformations it leads to are marked and characteristic. Around the affected joint

FIG. 7.*

spring up nodulated chalky masses, which every acute attack leaves behind, until hardly a joint of any limb but is not deformed and rendered unsightly, and to a large extent unserviceable by the disorder.´ (See Fig. 7.)

There are of course other diseases affecting joints, than either gout or rheumatism, and resulting in quite as much

* Extreme deformity of hand in chronic gout (after Garrod).

destruction of the joint itself and distortion of the limb. To name and illustrate only one of them. The hip-joint is frequently attacked with inflammation of a particular kind in children of what is called strumous constitution. In its early stage, this disease shows itself by the attitude of .the limb almost as much as by any subjective pain. The child fears to put its foot down, and stands with the affected leg raised and turned inwards, so that only the tips of the toes rest on the ground ; and actually there may be some shortening of the limb, whilst if the child be laid on its face, the normal fold of the buttock on that side will be almost if not entirely obliterated, the great protuberance itself being flattened out as compared with the opposite buttock. If the disease run its course, all these alterations become more marked, and the limb actually grow shorter, from disappearance of the head of the bone, and finally even firm union will take place between the hip and thigh bones. Then in walking the thigh moves with the hip, the body is bent, and one cause of spinal curvature is established.

" Housemaid's knee " is a swelling produced by enlargement of a sac containing fluid which exists beneath the skin in front of the knee-cap. Its enlargement is brought about by constant kneeling on hard surfaces without any protection pad, so that there is much pressure upon this part. Small elastic and occasionally painful swellings sometimes form at the back of the wrist from enlargement of similar sacs, which occur here in connection with the cords or tendons passing over the bones from the muscles which mar the fingers.

CHAPTER IV.

CHANGES IN ORGANS.

WE now come to a very important part of the subject, the organs of the body, and the influence their various structural changes exert upon external parts.   It will be well to pass with brevity over the natural disposition of the viscera, so that we may the more readily appreciate these facts.

The brain which is contained within the cavity of the skull is not closely applied to the sides of the cavity, but is invested with certain membranes, one thin and full of blood-vessels closely applied over the whole surface of the brain, the other dense and tough, lines the interior of the skull, but is not in actual contact with the brain. For the smooth outer surface of the one membrane and the smooth inner surface of the other are separated by a small quantity of fluid, and although here and there as age advances the virtual cavity which exists between the two membranes may become obliterated by the adhesion of the membranes, yet there is seldom complete blending unless from disease.

The cavity of the thorax, which extends from the root of the neck to the lower end of the breast-bone in front, is separated from the cavity of the abdomen by the midriff or diaphragm, an expanded muscular structure, which is dome-shaped, and on contracting tends to descend and thus to increase the capacity of the thorax. This occurs every time the lungs are filled with air in the process of respiration, the descent of the diaphragm being one main cause of the enlargement of the lungs so as to allow them to receive more air. At the same time by virtue of the muscular structures attached to them, the ribs are raised and turned outwards, and thus the thorax capacity is increased in a lateral direction. The lungs are each of them contained in a thin membrane or pleura, composed of two layers, the outer of which lines the wall of the cavity of the chest, and the inner is reflected around the air-tubes or bronchi, and blood-vessels entering the lung to cover the whole surface of the lung itself. Although in the normal state there is no actual cavity between the lung and the chest-wall, but only the virtual space between the smooth surface of the costal pleura (as it is called) on the one hand and the pulmonary pleura on the other, it is usual to speak of the " pleural cavity " when we mean that half of the chest cavity which contains a lung. The shape of this is roughly that of half a cone, the apex being upwards behind the highest rib and collar-bone, the base being on the surface of the diaphragm. But between the two pleural sacs, and encroaching chiefly on the space occupied by the left, lies the heart, contained in its own bag, the pericardium, and the large blood-vessels coming off from the heart.

In the abdomen are contained the organs of digestion and of the urinary excretion, as well as an organ which probably has something to do with the formation of the blood—the spleen. The spleen lies behind the last ribs on the left side in close contact with the diaphragm; on the right side behind the ribs, and in the arch of the diaphragm, extending also beyond the middle line to the left, and projecting a short distance below the ribs, is the liver. Partially concealed by its left lobe lies the stomach, the left extremity of which is in contact with the spleen. Immediately below and behind the stomach is the pancreas or sweetbread, which pours its secretion into the commencement of the intestine at a spot where the duct of the liver also sends the secretion of that organ, the bile. The rest of the abdominal cavity is occupied by the coils of the intestine, and the kidneys, the latter lying deeply imbedded in each flank. Lastly, passing down from the thorax through the cavity, and in close contact with the spine, is the great blood-vessel, the aorta, sending off branches in its course to the organs, and at the lower lumbar region dividing into vessels

4

for the supply of the pelvis and lower limbs. This great arterial trunk, taking blood from the heart, is accompanied in its course by a large venous trunk, the vena cava, bringing back the blood from the lower extremities to the heart. Much of the blood from the viscera passes by another route to the heart, viz., through the liver.

Confined as it is within a rigid box, it is not likely that the brain by any change in its shape can in the adult exert any influence on the shape of the skull and therefore of the external configuration of the head. So it happens that when any growth or tumour occurs in connexion with the brain, it does not cause the skull to change its shape, but rather the brain itself, which, being of less resistance, yields, and is displaced by the new growth. It is different, however, in infancy, when, as we have seen, the bones of the skull are but imperfectly developed, when their ossification is far from complete, and when, soft in themselves, they are separated by still softer and more yielding membrane. At this period of life if the brain enlarge it causes enlargement of the skull, and the condition which best shows that is that unfortunate condition popularly termed "water on the brain," or technically "hydrocephalus." In this state the brain is actually enlarged by the secretion of a large excess of fluid in the spaces, or as they are termed the ventricles, which exist within the nervous substance.

In proportion as the fluid is formed the capacity of these ventricles increases, and the whole bulk of the brain is enlarged. Then also the cavity' of the skull enlarges and the head assumes a rounded shape, in which the cranial part is out of all proportion to the small facial part in a painfully grotesque manner (Fig. 8). A comparison of this form of enlargement of the head with that pro-duced by the perverted growth of the cranium in rickets (Fig. 9) shows that in hydrocephalus the skull tends to the globular form, whereas in rickets, the square-shaped protuberant forehead and the projecting occiput are surmounted by almost a flattened vertex. Sometimes, in consequence of this hydrocephalus, the cranium attains an enormous size. Nor does it always, although very frequently, terminate in early death. The progress may be arrested, but the enlargement never subsides, and throughout life the rounded form of the head may be retained. Often bones not present in the normal skull are developed to fill up the spaces of membrane.

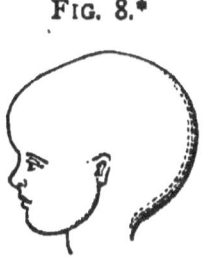

FIG. 8.*

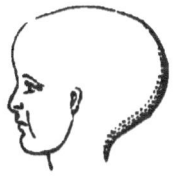

FIG. 9.†

Here we may turn aside for a moment to glance at

* Outline of hydrocephalic head.
† Outline of ricketty head.   The deformity in this instance is not very pronounced.

the position assumed by the phrenologist. It is a position wholly untenable, and we should not have deemed it worthy of refutation here were it not that misled by the recent discoveries of physiologists with regard to certain facts of experiment and disease, it has been reiterated that not only have certain regions of the brain their special functions, but that these regions are indicated on the surface of the skull by prominences which denote similar development of the brain within. Three distinct facts converge to utterly explode the phrenologist's standpoint. The first is that the brain is not in actual contact with the skull, but is separated from it by a membranous sac between the inner wall of which and the surface of the brain there is not any union. The second is that the inner surface of the skull does not correspond at all in configuration with the outer surface, for on the vault of the skull there is interposed between what are called the inner and outer tables a layer of loose textured bone of variable thickness, and in the frontal region there is developed to a very variable extent in different individuals a series of cavities between these two tables, the size of which amplifies or diminishes the amount of the forehead. The third is that even in parts where these cells do not exist, the thickness of the skull is very variable in different individuals, and prominences occur in larger number and extent in some than in others, according to the size and strength of the muscles attached to them. Thus a small skull with thin walls may contain a comparatively large brain, and *vice versâ*. But even if the reverse were the fact, mere size of brain goes for nothing in estimating its capacity, and more will be learnt by studying the complexity of its constituent elements than any amount of happy guesses and shrewd surmises arrived at from a study of the external surface of the skull, especially when it is remembered that all the inferior parts of the brain are concealed and have no relation to the outer regions of the skull.

In the neck an irregular enlargement on each side results from enlargement of bodies connected with the lymphatic system, and called lymphatic glands, which, forming a chain on each side of the neck, as well as beneath the jaw, may inflame and lead to the formation of abscesses, which, after bursting and discharging their contents, leave behind unsightly puckered scars in the skin. The tendency to the spontaneous enlargement of these structures, and their ready inflammation, is most marked in the strumous or scrofulous type of constitution. Sometimes an enlargement of these neck-glands is shared in by similar change in the like glands in other parts of the body, and large irregular tumours may grow up and conceal the natural shape of the neck. The Derbyshire neck is the result of the enlargement of a structure the precise office of which is unknown. This is a fleshy organ called the thyroid body situated immediately below the prominence in part of the neck, known by the name of " Adam's apple," and consisting of two lobes connected by a narrow isthmus, which passes across the air-passage, the lobes lying one on each side of the voice organ or larynx. By its enlargement, which often is excessive, there ensues a large prominence in front of the neck, this is the wen or " goitre," frequent in Derbyshire and in many parts of Switzerland and France.

The size of the chest, as we have said, is ever changing. It enlarges as the lungs expand, and contracts as they empty. Its form and dimensions also largely depend upon the condition of these organs. If they remain from birth not fully inflated, there is corresponding smallness in the form of the chest itself. If on the other hand they are of large size, and unusually capacious, the thorax is large, broad, and deep. The change in shape which the chest undergoes during respiration is as follows: with inspiration the diaphragm descends, that is to say, becomes less convex towards the chest; and the lateral walls of the chest are enlarged by the ribs being raised by means of certain muscles passing between them, and the upper ones by muscles going from the spine in the neck to the upper rib, and also slightly everted, this latter action being most marked in the lower ribs. An elevatory movement combined with one of expansion takes place every time we take a breath, and if the breathing be forced and hurried, other and more powerful muscles which have attachments to ribs, act on these bones and increase the area of their movement. But as the ribs are (the six upper directly and the three following indirectly) connected with the sternum or breast bone, this bone is, at the same time as the ribs are raised, thrust bodily forward. There is thus an enlargement of all the diameters of the chest, the vertical, the transverse, and the antero-posterior; the cavity is increased in length, in width, and in depth from before backwards. When, on the other hand, the lungs are completely emptied of air, that is, at the close of a number of forced expirations, the chest is narrowed in all those diameters, and its circumference is proportionately less. Contrasting the position of the ribs in the two conditions, note that at the end of a full inspiration they have a nearly horizontal direction, whereas at the close of a deep expiration they are sloped downwards and forwards, and are therefore more vertical. So that, in fact, the chest looks longer but narrower when its lungs are less expanded than when they are fully distended. Now it is not at all uncommon to meet with people the conformation of whose chest approaches what may be called the " inspiratory type," and quite, if not more common those in which it is rather of " expiratory type." The small, contracted, long and narrow chest is sadly too common. By no means indicative of present disease, it yet shows a lessened lung capacity more plainly than any other surface indication, and is therefore to be regarded as a warning sign. What produces this expiratory form of chest? It may be assumed to be due to no single condition, but rather the grafting of faults of training, faults of dress, superadded to those primary defects over which the possessor has no control, and for which heredity is responsible, and the artificial existence of the race more responsible still. The preventable causes ought to be removed, and all should be done that is possible to remedy or lessen the unpreventable. A multitude of errors in our artificial, over-cultured modern life, ought to be, nay, must be, cleared away; and stricter attention paid to the laws of health if the tendency to this defect is to be combatted at all. That it ever can be thoroughly and completely overcome, would, however, seem almost impossible, seeing that so many of the essentials of civilisation depend upon the sacrifice of health; and numberless trades and occupations of themselves favour the perpetuation of the defect. Sedentary occupations, especially such as require constant bending to work; the evil produced by close study over books on benches which afford no support for the back, and the want of complete and systematic exercise of muscles and of lungs, tend to produce

a day, so does the shape of the abdomen alter. On this point we may only state here, that food which is liable to much fermentation, which is not readily digested, and causes the development of much gas, acts chiefly in causing distension of these organs. Hence it is that vegetable feeders have large bellies. Continued and often-repeated temporary causes may thus lead to permanent deformity.

Any of the solid organs in the abdominal cavity may from various causes undergo change in shape, and size, and lead to departure from the normal form of the abdomen. This may be from an enlargement of the organ in all directions, or from its unequal enlargement from the growth of tumours in it. Swellings in the lower part of the abdomen or in the groin, may be produced by "rupture," in which a portion of the contents of the abdomen is protruded through the abdominal wall where it is weakest. Such swellings will increase under efforts which bring the abdominal muscles into action, as for example, when the breath is held and a vigorous attempt made to lift a heavy body.

But there is one condition in which the abdomen is uniformly enlarged and assumes a globular form. That is, dropsy of the abdominal cavity, where fluid derived from the blood fills and expands the sac in which the viscera are contained. In proportion as the amount of fluid is great, so will the form of the abdomen tend to become more spherical ; for when it is slight the fluid lies at the lowest part, and the contour will vary with the position of the subject—the intestines floating on the surface of the fluid.   But when this reaches the maximum amount it tests the expansile capability of the abdominal walls to the uttermost, and they become stretched tightly and the abdomen gains a spheroidal shape.

CHAPTER V.

ARTIFICIAL ALTERATIONS IN SHAPE.

It is the prerogative of civilised man to be the slave of custom and of fashion.   Still among savage races there is often as much adhesion to absurd, irrational, and almost harmful observances as among the so-called cultivated.   Setting aside such deformations as have no great significance, we may speak of two barbaric customs only, and then turn to civilised communities and see the analogous instances of ignorant folly, ten times more foolish and stupid (because there they have not the excuse of ignorance), that prevail among them. It is customary among many American races, especially the Peruvians, to subject their infants to a process of compression of the skull, which is carried out to such

an extent, that a deformity remains throughout life,
and the frontal region instead of rising boldly up above
the orbits appears as if cut away, and as if a long
but slightly inclined plane were left rising gradually from
the orbit to the middle of the vertex of the skull. The
resulting shape is so peculiar, and the facial angle thus
produced so small and brute-like that there is no wonder
that it should at one time have been regarded as typify-
ing a peculiar race of man. The process, however,
has been now frequently observed and recorded by com-
petent travellers, and it has been shown that in different
tribes various kinds of compression are practised, with
the result of procuring curiously unsymmetrical crania.
The accompanying figure (Fig. 11) is a reduced outline
from an engraving of such a skull
from Titicaca in Peru, in Dr.
Prichard's 'Natural History of
Man,' and the author quotes the
opinion of Dr. Scouler as to the
manner in which the compression
is effected. He says, "The pro-

FIG. 11.*

cess is slow and gentle, so that the child does not appear
to suffer in any way from so unnatural a process, nor
do the intellectual qualities of the individual appear
to be in any degree affected by it; on the contrary, a

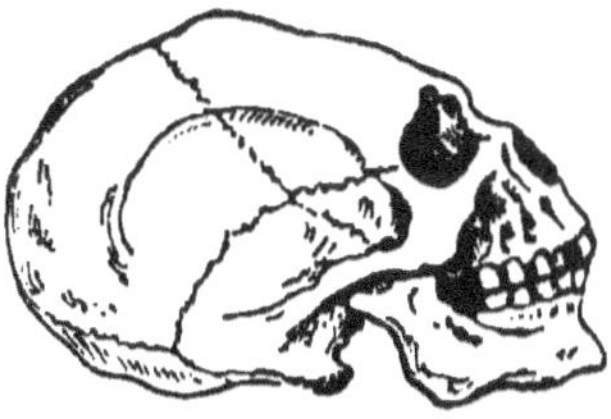

* Compressed cranium of a native of Titicaca.

flat head is esteemed an honour, and distinguishes the freeman from the slave."

The Chinese custom of cramping the feet of female children, consists in so binding them up that the toes and heel are almost brought into contact, and the sole of the foot bent upwards in a high arch. The result

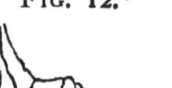

FIG. 12.* upon the shape and structure of the foot and of the leg is very marked (Fig. 12), the bones of the toes and foot itself being small and atrophied and compressed, whilst the muscles in the leg which should act on the toes, but which are never called into action, are wasted and small.

Turning to the "civilised" customs; the high-heeled boot of modern female fashion does not favour the development of calf-muscles, seeing that it throws the foot forwards on the toes, and normal progression cannot take place. The wearing of small and tight boots—a far more prevalent evil—leads, as is well known, to much discomfort in progression and much deformity of the feet. The toes cramped, distorted, and displaced, become the seat of thickenings of the skin where pressure is most felt, and corns and bunions come to disfigure the natural elegance of a well-formed foot. One other evil resulting from the wearing of improperly-made boots is the flat

* Foot of Chinese woman (after Humphry).

foot, in which the strong fibrous ligament that passes under the bones of the foot and maintains its natural arch is weakened, and the whole structure yields.

The wearing of tight stays and corsets—the folly of which is, that they are worn with the avowed object of distorting the figure—leads to compression of the lower part of the chest, interfering with its movements in breathing, and displacing the abdominal organs. Apart from the foolishness of the practice from an æsthetic point of view, it is still more unwise in a hygienic, as there can be no doubt that compression of the chest must both impair the action of the breathing muscles, and render the muscles compressed by it weak and ineffective. But in spite of much attack the practice holds its ground, and so it will as long as there are foolish women enough in the world to think that a small waist is a thing of beauty.

Other pieces of clothing which are worn habitually tighter than they need be, such as stockings, may not be productive of much change in figure, though to the last-mentioned articles must undoubtedly be attributed many cramped and deformed toes; that tight garters should be discarded is obvious, since, by compressing the veins of the leg, they retard the flow of blood from the limb, and are a prime cause of enlarged veins in the leg and swelling of the foot and limb.

5

The barbaric habit of wearing earrings might almost be considered a harmless one, for it does not much alter the shape of the ear-lobe, were it not that the act of piercing the lobe and inserting rings is apt to be followed, in young children especially, by troublesome inflammation of the skin at the part pierced. It is no uncommon thing to find, among the lower classes, mothers nursing infants a few months old who are already in the possession of the coveted ornament.

But it is time to turn from this section of our subject, so leaving the question of deviations of form (although much more might be said of them did space permit), we pass to those of changes in the colour of the body; and in the first place let us make clear upon what this colour depends.

———

CHAPTER VI.

. THE COLOUR OF THE HUMAN BODY.

WHAT is the cause of the red colour of the blood? It took a long time to answer this question. Men had to wait until the microscope was first applied to the investigation of the blood by Malpighi, about two hundred years ago, before it was answered. And then it was found that this blood—this red fluid was really, so far as its fluid part went, colourless, but that with the colourless fluid or serum were mingled a number of small round bodies to which the name corpuscles were given. Then it came to be found that of these corpuscles there were two kinds, some, and these formed the large majority, were of a pale yellowish red colour, hence called "red corpuscles;" whilst others, larger, less regular in shape, were pale or devoid of colour. Ever since that time attention has been largely directed to the nature and properties of these constituents of the blood, and we must describe them in slightly more ample detail, limiting our remarks only to the red corpuscles. If a drop of blood be placed on a glass slide and examined under the microscope, the field will be seen to be in great part occupied by small circular bodies, in which a faint yellowish tint indicates their possession of some colouring matter. They are not only circular, but are flattened from side to side, and careful observation will show that their flattened surface is really depressed (Fig. 13). They are in fact small discs with a slight concavity on each surface, a character which is perceived at once when a corpuscle happens to roll over whilst it is being observed, and the observer sees it edgeways. A single cor-

FIG. 13.*

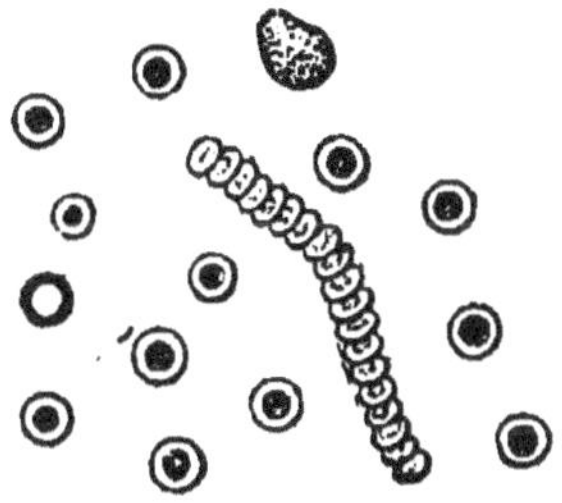

* Blood corpuscles, all with one exception being red.

puscle when viewed alone seems to have very little colour at all, a mere faint tinge, and if a drop of water be added to the blood the corpuscle will swell up and become globular, and its colour will become even fainter than before. But, even under the microscope, it will be seen how the corpuscles are sufficient to account for all the colour the blood possesses, for the little bodies have a great tendency to collect in groups, especially forming rows like a pile of coins, and then the colour of the whole group is more manifest. If further proof be needed that the colour of the blood is due to these little particles, it may be found by collecting a small quantity of blood and exposing it to the air. After a short time the fluid will set—it undergoes what is called coagulation—the resulting solid mass being called a clot. This clot is formed by the production of a substance from the fluid part of the blood known as fibrine, which does not exist as such in the blood when circulating through the body. The fibrine is formed in short and fine threads which entangle the corpuscles in their meshes. Now the red corpuscles being relatively heavier than the fluid part of the blood sink to the bottom of the vessel during the process of coagulation, and more of them will sink the slower this process takes place. Hence it comes to pass that whereas the lower part of a clot so formed is of a dark red colour, the upper layers may be almost colourless or only just tinged by reason of the entanglement of a few of the red corpuscles, which have not had time to sink to the bottom of the vessel before the clot set. A microscopical examination of a portion of the lower part of the clot will show that there the fibrine has entangled nothing but red corpuscles, whilst hardly any of these corpuscles will be found in the upper strata. Again it has been found that pale blood, or the blood of pale persons, contains fewer red corpuscles than the average quantity in health, and even quite recently it has been ascertained that the red corpuscles themselves may contain less colouring matter than the normal. The physiologist has long since succeeded in separating this colouring matter in bulk, and obtaining it in the crystalline state.

The chief constituent of the blood pigment or hœmoglobin, is carbon, which forms more than one-half of it; oxygen, nearly a quarter; nitrogen and hydrogen make up all but a small fraction of the remainder, which yields sulphur and iron. The most important property it possesses is the power of combining with oxygen, and when this combination occurs the colour of the blood changes ; it becomes brighter and more scarlet. If, however, the oxidised blood be exposed to carbonic oxide or carbonic acid, the colour changes to dark purple. This readiness to combine with oxygen and to give it up again appears indeed to be *the* essential property of the red corpuscles, and the difference in colour thus produced constitutes the essential difference between the blood which flows *from* the heart in the arteries to all parts of the body, and that which is returned *to* the heart from the tissues by the veins. When an artery is divided the blood that jets forth with every beat of the heart is of a bright red colour ; on the other hand, that which flows from a divided vein in a constant stream is dark and purple. The reason of this difference in colour between arterial and venous blood arises from the fact that before entering the left side of the heart to be sent through the body by the arteries, the blood passes through the lungs, and then the carbonic acid with which the venous blood is charged is at once exchanged for oxygen by the process of respiration. Again, the bright arterial blood carrying oxygen to the tissues, the muscles, glands, &c., receives carbonic acid in exchange for its oxygen, which goes to maintain the tissues in their constant process of change, and the dark carbon-laded blood courses back to the heart to be again renewed in the lungs. If from any cause this process of oxygenation of the blood be carried on imperfectly, or if the heart be unable to do its work efficiently, so that the venous blood does not pass on with sufficient rapidity, then the blood in the tissues is less bright, and as will be seen later on, the surface of the body denotes the darkening.

Although much of the colour of the body depends upon the -blood, and although some of the chief indications of derangement of health may be found in changes of the colour due to blood conditions, yet there is also throughout the body a large quantity of special pigment deposited in different regions. This colouring matter wherever it occurs is brown or black, and is met with in the form of minute particles or granules within the constituent cells which make up the various structures where it is found. Derived as it probably is from the blood-colouring matter, its composition is clearly allied to this substance, only it is even richer in carbon. In some places where it is deposited it subserves definite uses; but for our present purpose we have only to consider it as affecting the colour of the body, and we are thus limited to its consideration in influencing the colour of the skin, of the hair, and of the eyes, or, more strictly speaking, of the "irides."

Regarding then, for our present purpose, the skin only as the seat of colour, and leaving for others* to deal with its more important functions as an excretory organ of the body, it will suffice to describe the structure of this integument just so far as is necessary to give a clear conception of the cause of its colouration.

* For full details as to structure of skin, see work in this series on the Skin.

The skin, which serves as the protective covering of the body, is a tough membranous structure, consisting of three parts, an external layer of *epidermis* or scarf-skin, an internal, sensitive and vascular layer called *corium* or true skin, and an intervening soft, spongy layer called the *rete mucosum*, or, after its discoverer, *rete Malpighii*. This middle layer is the seat of collections of fine pigment granules, which are the primary cause of diversity of colour. In dark persons this layer is thicker, more spongy, and contains more pigment than in fair ones, whilst in the Ethiopians it is very thick and very black, so that the colour of the blood in the underlying corium is never seen at all. There are some diseases, one especially, and that of rare occurrence, which are marked by a gradually increasing deposit of this pigment, so that a previously fair skin may become gradually more and more bronzed,* the regions which are naturally the seat of most pigment being those which suffer most.

A "bruise" is the name given to the effect produced on the skin by an injury which does not suffice to break the skin but only to contuse it. The blood escaping from such blood-vessels as are torn in the injury, collects under the skin, and the gradual variation in tint that a

* So marked is this that the affection in question has been styled "Bronzed-skin Disease," although the coloration of the skin is only a part, and that a small part, of a deep-seated disorder.

bruise undergoes, from purplish-red to green, orange, and yellow, is due to the changes taking place in the colouring matter of the effused blood, whilst this is being very gradually absorbed. All schoolboys must have enjoyed opportunities of observing these colour-changes in the familiar " black eye."

Tattooing consists in the insertion into the true skin of some pigment as charcoal, gunpowder, &c., by means of punctures made in the skin and the rubbing in of the particles. Once inserted there the pigment remains, and the tattoo-mark cannot be removed by any means short of excision. The accompanying sketch (Fig. 14) is from a microscopical specimen of a piece of skin from a highly-tattooed arm. Observe that the pigment is not contained in the epidermal cells, but is collected in little heaps and masses between the bundles of tissue that form the dense cutis. To remove them by simply rubbing the cuticle off, or raising a blister, will not suffice ; and as a considerable amount of tissue would have to be taken away, a scar must ensue. Hence it may be laid down as a rule, without exception, that no tattoo-mark can be removed without leaving a scar to indicate the place it formerly occupied.

The whiteness of scars depends on the fact that the middle layer of the skin (or at any rate those constituents of it which contain pigment), is not reproduced with the

healing of the wound ; and, further, the new tissue formed
in the corium contains fewer blood-vessels than the old
tissue which it replaces.   A curious but interesting fact
in connexion with scars is their growth with the body ;
that is to say, if a wound or injury sufficient to leave a
scar be inflicted in early life, the scar formed will retain

FIG. 14.*

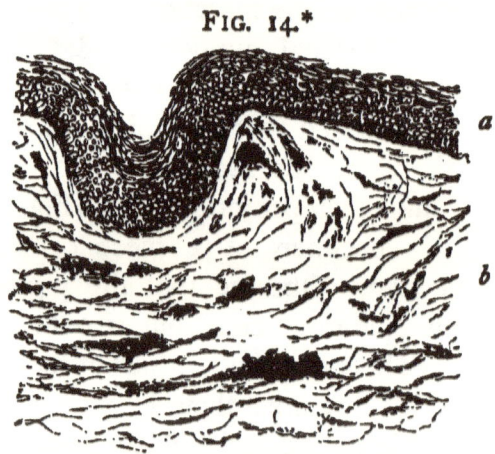

its comparative size to the rest of the body, although this
may have become three or four times as large as at the
time of the infliction of the wound.

The *nails* and *hairs* are structures, which, in truth, are

* Microscopical section of tattooed skin : *a* is the surface layer or
epidermis, in the deeper layers of which the true pigment of the
skin is lodged ; *b* is the cutis, and imbedded in its texture will be
seen black masses of the tattoo-pigment.

nothing more than modifications of epidermis. The former consist of a transparent, horny substance, firmly imbedded in the true skin, the pink colour of which can be seen through them. Constantly growing by additions at their posterior ends they project over the ends of the fingers and toes, and if uncut may attain extraordinary dimensions. The nails are never coloured by the natural pigment of the skin. They are narrow, long, sharply-curved and pointed on the delicate fingers of the finely-formed, clear-complexioned inheritant of a tendency to consumption; whilst after acute illness they frequently show ragged transverse lines and fissures denoting that nutrition has been gravely affected. The *hairs* which exist over nearly the whole of the body and attain their greatest development on the scalp, are horny, fibrous-looking structures which spring from little pits or follicles imbedded in the true skin. The colour of the hair depends upon the accumulation of fine granules of pigment in its substance.

In the accompanying sketch (Fig. 15) will be seen a minute portion of three varieties of hair from the heads of different individuals. One (*a*) is a portion of a white hair from the head of an old man; the fibrous and scaly character is well seen. It is absolutely devoid of pigment in any part. Then comes a piece of a light yellow hair (*b*), and the main difference between this and the preceding will be seen in the presence in the former of dots and linear streaks of pigment which had an amber colour, although of course they appear dark in the engraving. Lastly we have a black hair (*c*), and this will be seen studded with pigment granules in all parts. We learn from this then that the colour of the hair depends on the presence of pigment particles more or less widely dissemi-

FIG. 15.*

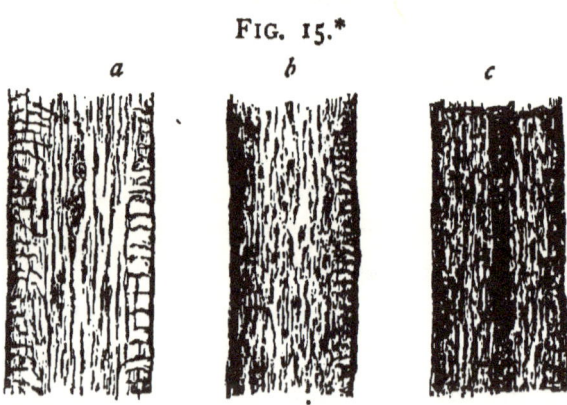

nated throughout its substance, whilst a purely white hair is distinguished by the negative character of the absence of all such pigment. That men's hairs have been known to turn " white in a single night," through sudden fear, or some equally powerful emotion, is perfectly true. Several well-authenticated instances are recorded. Not that the pigment actually disappears from the hair ; it is probably

* Hairs of different shades of colour.

only concealed by the development of air or gas which suddenly takes place at the roots of the hair and penetrates the structure. Nor do such changed hairs always remain invariably white; as they grow, the pigment may still be formed, so that the same hair may become half white and half black.

Before leaving the subject of the colouration of the skin, one word must be said as to the colour of those structures called the mucous membranes. Where the skin is said to end—that is, at the orifices of the body, at the eyelids, at the ears, nose, mouth, &c.—it is replaced by a membrane which in its minute structure differs but little from the skin itself. Like the skin it consists of several layers, but the colour of these mucous membranes is red or pinkish red, the very vascular tissue beneath showing through the thin membrane. Hence it is that the lips are red, and the same "vascularity" prevails throughout all the canals and passages into which these orifices open.

Finally, no description of the colour of the body would be complete without reference to the colour of the eyes. Of the general structure of the eyeball and its constituent parts it does not concern us to speak here.* The anterior part of the organ is all that is visible to us. We see, between the eyelids, a central convex, glassy struc-

* For more ample details see the work in this series on the Eye.

ture set, as it were, in front of a white or bluish-white surface, " the white of the eye," but little of which is visible between the lids when the eye is directed straight forwards. This white part is well furnished with small blood-vessels, which when the eye is rubbed or irritated by particles of dust, &c., become visible and give it a reddened or " blood-shot" aspect. The round aperture seen behind the centre of the glassy structure and called the pupil appears black to the observer; it is bounded by a circular membranous or rather muscular curtain called the *iris*, and the pupil can increase in size or diminish by reason of the muscular fibres which compose the chief thickness of the iris; fibres going round it, being for the purpose of diminishing the size of the pupil, and others radiating on all sides, for the purpose of increasing its size. It is this membranous curtain that is coloured, and gives the colour to the eye, that colour depending upon collections of pigment granules in a layer of cells which covers its hinder surface. In dark brown eyes this pigment is very abundant; in " blue" and " grey" eyes it is very scanty. There is also a large amount of pigment in the cells composing the middle membrane of the globe of the eye, the choroid coat as it is called. The choroid is really an extension of the membrane forming the iris. In dark-haired and dark-complexioned people this choroidal pigment is very abundant, in fair people it is less abundant, and in the curious type of "albinoes" there is no pigment at all in the choroid. As this choroidal pigment subserves a definite use in vision, causing the absorption of extraneous rays, such people are unable to see in strong direct light, and consequently in the bright light, or even in the simple diffused light of day, they are obliged to keep their eyes half closed in order to see anything.

The glassy cornea sometimes becomes opaque, or the seat of opaque white patches from inflammation. The glazing of the eye in death is produced by the shrinking of this same structure. Another change affecting the cornea, as old age approaches, is the formation of an opaque white line around the margin of the cornea, and travelling round it, so that it does not at first make a complete circle; it is known as the *arcus senilis*. It is part and parcel of the degeneration of tissues which is the natural accompaniment of age. Another old-age condition, which sometimes takes place, is the hardening and opacity of the crystalline lens—known as cataract, such opacity being visible to the observer through the pupil.

Having just mentioned the case of "albinoes" it may be convenient to sum up their characters now and have done with them. A true "albino" is one whose skin, hair and eyes are wholly devoid of pigment from birth onwards. The skin is thin and fair, the delicate flesh-tints due to the colour of the blood showing through the cuticle are very manifest; the hair is a pure white, on the scalp, the eyebrows, eyelashes, &c. ; the pupil looks red, like a white rabbit's eye, from the absence of any pigment on the choroid at the back of the eyeball, so that the colour of the blood in that vascular membrane is reflected through the transparent media of the eye; and the iris also is destitute of colour, or rather its colour is pinkish or red for the same reason as the choroid is. This condition of "albinism" is said to be less common among the white races than among the black, so that a "white negro" is a verity. There are instances also where this albinism instead of being general is partial, the result being a sort of piebald.

Among the white races, the Indo-European stock, there are two well-marked types of colouration of the body. From each of these there are numerous departures and variations, still even in common parlance it is customary to speak of a person being "fair" or "dark" according as he falls under one or the other description. The typical fair-complexioned blonde has a delicate soft skin, through which the emotional disturbances of the circulation are readily seen, fine yellow flaxen hair, and irides of a bluish or greyish tint. The dark-complexioned or brunette has a coarser skin, with more pigment in it in the usual seats of this pigment—black hair, or shades of brown verging on black, and irides of a nut-brown colour.

The difference in coloration of the skin between inhabitants of hot climates and those of temperate races, is probably due to an exaggeration of the conditions which are undergone by white-skinned men after prolonged exposure to heat.   The cause of the colour of the negro, rests in the great accumulation of pigment in the deep layers of the epidermis ; that pigment is derived from the colouring matter of the blood, and the reason of its excessive production perhaps depends upon some deeper seated changes in blood formation about which we can only form conjectures.   It is suggested that the liver, which is the organ most readily influenced by changes in temperature, and which secretes a product largely pigmented, viz., the bile, may have its function so modified that an excess of pigment is, as it were, left in the blood.   At any rate there is a reason for the coloration of dark skins, since they serve as a protection from the effects of heat to the delicate structures beneath.

The "tanning" of the skin of exposed parts, which occurs after a few days' exposure to the air and solar rays, is due doubtless in part to increased vascularity and escape of pigment from the blood, just as any irritant applied to the skin will produce pigmentation from the congestion it causes.   Other and deeper influences may

6

be at work, but we can hardly invoke here the hypothesis that derangement of hepatic function, gravely as this is affected by heat, has anything to do with it, seeing that the result is only shown upon those parts of the skin which are exposed to the air. The influence of light in this connexion should not be lost sight of; for those who work in dark places are pallid, just as plants kept in the dark are; and though we may not be able to explain the *modus operandi*, yet we must think that light has some effect. Fair skins are especially liable to pigmentation from exposure to air and light, as is seen in the production of " freckles " or " sun-spots " on the face and neck, and hands. These are nothing more than limited collections of pigment, and as is well known they often disappear when youth is passed.

CHAPTER IX.

CHANGES IN COLOUR IN HEALTH.

HITHERTO we have spoken only of changes in colour due to alterations in the amount of pigment in different regions of the body. Many of those changes are of but slight importance, for depending upon local influences they have no direct relation to the general health of the individual. It is far otherwise with the colour changes we have now to deal with; for although, as we have before pointed out, it is probable that all pigment owes its primary origin to the colouring matter of the red blood corpuscles, yet variations in degree of pigmentation of parts are not necessarily associated with any marked change in the blood itself. It will be remembered also that the colour of the skin depends upon two factors, the chief being that of the blood circulating in the deep layers, that gives the "flesh-tint" which artists aim so strenuously to reproduce in all its delicate and varied hues—a colour too, the intensity and delicacy of which must largely depend on thickness of cuticle ; and the other being the presence of pigment in the cuticle—a very varying condition, but one which influences largely the predominant colour, and which, if in excess, as in the negro, will suffice to completely disguise the flesh-tint. However, even in the negro there are still regions which show the blood colour, e.g., the mucous membranes ; and observation of the colour of the lips, of the nails, or of the conjunctiva of the eye, will reveal more accurately than any change that may be noticed in the skin, the change that has taken place in the vascular regions beneath.

The difference in colour produced by difference in the quantity of blood in a part may be shown by a very simple experiment. If one hand be held above the head whilst the other is allowed to hang down by the side for a few minutes, and then the two hands compared, the one will look pale and white, the other flushed and red. The first is white, because by holding it up the return of blood from it by the veins has been aided by the action of gravity; in the other it has been retarded for the same reason. Thus temporarily there has been produced a condition of comparative bloodlessness or *anæmia* in the one hand, and a condition of plethora or congestion (*hyperæmia*) in the vessels of the other. The experiment may be varied by pressing out the blood from one hand, when you will get a white spot where the pressure was applied, and conversely, if the part be rubbed violently, it becomes red and warm. The reason of the pallor in the first instance is obvious enough, the pressure squeezes the blood out of the vessels of the part. The reason of the redness in the latter example is not so obvious, and requires a little further explanation. All the blood-vessels, the arteries more particularly, have muscular walls, which can contract when the nerves which go to them are irritated, just as the muscles of the arms contract if the nerves going to them be stimulated, as they always are when by an act of the will the brain sends its message down them to exert a particular movement. The *modus operandi* however in connection with the blood-vessels is different from that of the muscles of the arm, or any of those which are under the control of the will; for the muscular walls of the arteries are not under voluntary control. The nerves which go to them do not pass directly from the brain or spinal cord, but are connected with a part of the nervous system, which, although in part dependent on the cerebro-spinal, is distinct enough, both in its arrangement and structure, as well as in its functions. This is the system of organic nerves, or, as it is also called, the sympathetic system, and it consists essentially in collections of nervous cells called ganglia, connected by bundles of nerve fibres down each side of the spine, and from which proceed other nerves to the involuntary muscles all over the body; to the muscle of the heart, to those of the intestines and stomach, &c., the movements of all of which are not under voluntary control, but are essential to the performance of the function of these organs. The nervous system is then in its central parts duplex; there is the brain and spinal cord from which nerves of voluntary motion and sensation arise, and there is the sympathetic system from which nerves of organic life, of involuntary movement arise. Going back to our illustration, we find that the arteries of the body are kept in a state of contraction by the sympathetic nerves. It has been shown that when the sympathetic nerve in the neck of a rabbit is paralysed, the ear on that side becomes red and hot, and increases in size. There is in fact an increased flow of blood to it, and this flow is due to greater enlargement of the blood-vessels, i.e., they are dilated. The influence of the sympathetic nerve has been taken off their contracting walls, and just as a limb is paralysed when its nerves are injured, the coats of the blood-vessels are paralysed. But when the nerve is stimulated, the vessels contract, and the previously engorged ear becomes cold, pale, and white, less blood passing through it. And now we come to another fact which must be accepted before we can explain the reason why the hand gets red when it is vigorously rubbed. It is this: although the cerebrospinal nervous system has no direct influence over the involuntary muscles supplied by the sympathetic nerve, it has an indirect influence. It possesses the power of arresting, or diminishing, or, as it is called in scientific phrase, "inhibiting" the action of the sympathetic nerves. Now the act of rubbing the hand stimulates the fine sensory filaments of the sensory nerves in the skin; their stimulation exciting them to increased function, they inhibit the action of the sympathetic nerves governing the blood-vessels in that part; and the control of the sympathetic being temporarily suspended, the arteries dilate, and becoming capable of receiving more blood than usual, the part they supply becomes temporarily red and hot. The same explanation holds for the effects of other kinds of irritation or stimulation of sensory nerves, e.g., the red blush produced by the application of a mustard poultice, and carried further still, the first effect of a blister is to cause reddening of the skin, and later as the blister rises exudation of fluid from the blood in the dilated vessels of the cutis. In speaking of pigmentation of the skin due to the effect of heat, we said that the first effect was an increase in the flow of blood, and no doubt it is due to this that there comes to be increased formation of pigment. For a brown stain is often long left after the application of mustard or a blister, or after the skin has been inflamed for some time.

We are now in a position to explain the phenomena of *blushing* and of *pallor* of the surface from fright. In blushing, which is always most marked in thin-skinned, fair people, and which is by no means limited to the face, but may extend over the neck, shoulders, and chest, the arteries are temporarily dilated. The sense of shame acts through the medium of the brain in an involuntary manner, and there is a temporary arrest of the function of the nerves going to the blood-vessels. The blood seems to "rush" to the cheeks, and the individual feels hot and uncomfortable, whilst to the observer the rosy flush will be seen mantling the cheeks, ears, and forehead, and reaching to the roots of the hair. The heart beats quicker, and the whole circulation is temporarily excited. Then as the emotion sinks, the blood tint fades away, and matters return to their previous state of quiescence.

In the opposite condition, that of pallor, which is associated with the emotion of fear, especially sudden shock, the skin becomes pale, the lips lose their brightness, and often a sense of coldness even to shivering is felt. Here we have a temporary diminution in the flow of blood—the brain acting upon the heart—and the surface being less supplied with blood. A similar effect is produced in fainting. The rationale of the fainting state is a temporary failure in the action of the heart, and the brain failing to receive its full supply of blood, unconsciousness supervenes. The bloodlessness of the surface typifies the bloodlessness of the brain. The return of blood to all these parts, and the restoration from the fainting state is accompanied by a sense of heat and fulness.

And so again the final hue of death, in all time obvious and striking—"pallida mors"—is due to the final beats of the heart, and dying contractions of the arteries sending the life-stream slowly on its way to stagnate in the veins. The emptiness of the arteries and the capillaries gives a pallor to the surface, which the laden veins do not overcome.

CHAPTER VIII.

CHANGES IN COLOUR IN DISEASE.

PASSING from conditions giving rise to pallor, which are physiological, and therefore not departures from healthy action at all, we are confronted with a set of causes of undue paleness of the surface which depend on more or less grave alterations in the constitution of the blood. The pallor in such cases is general and more or less lasting. It may be due to a general diminution in the quantity of the blood, with or without any alteration in its quality. Great losses of blood are obvious causes of this anæmia, and in such cases it comes on rapidly. Or a long and tedious illness gradually interfering with the nutrition of the body, and therefore with the formation of the blood, will be denoted on the surface by the paleness of the skin dependent on the diminished and altered condition of the fluid in the blood-vessels.

More common than all, and striking enough in its effects, is the pallor produced by unhealthy trades and occupations ; by life in badly-ventilated, ill-lighted, and over-heated rooms. Then just as the plant grows pale when kept from the light, so the man or woman loses the tint of health and presents a blanched appearance. This is the state of general bloodlessness or anæmia which is so prevalent in so many of our townsfolk and factory hands. The lips lose their natural redness and become pale, as do the gums also, and the other visible mucous membranes. A white line on the thin part of the gum, where this stretched over the prominent canine teeth, is an indication of this anæmia. The skin is white and no longer has the flesh tint natural to it. Many troubles arise out of this condition of impoverished blood; for, being poor in corpuscles and deficient in colouring matter, the functions which these subserve are impaired, and shortness of breath on exertion, caused by the demand of the tissues for oxygen being greater than the small number of oxygen carriers can supply, disturbed and imperfect digestion, disorders of the mental faculties, sensorial derangements, &c., all follow in the train.

But it may be said that numbers of persons are naturally pale, and apparently bloodless, but yet enjoy good health, with none of these grave discomforts. That is true, but it is only true because their pallor is not the pallor of true anæmia; it may be dependent upon the condition of the circulation in their skin; or the skin may be thicker in them than in other folk, who, with no more blood, are yet red-faced and healthy-looking. Then these people do not exhibit the unnatural paleness of the lips or nails, or the mucous membrane under the eyelids— the only places where in dark-complexioned thick-skinned people the objective signs of anæmia can be with confidence ascertained.

As to this state of anæmia, all its causation may be summed up in derangement of those processes which are concerned in blood-making; of these, digestion and oxygenation are the chief; and of the causes, consequently, ill digestion and bad oxygenation, in other words improper food and bad air, are the predominating circumstances contributing to this state.   Various as are the diseases of which anæmia plays a part, the production of the anæmia itself always comes back to this.

Apart from its association with any grave disease this poorness in the quality of the blood is common among young women; especially those who, accustomed to outdoor life, are called upon to work from morning to night within the house—and often in them the pallor of the surface is not merely paleness, but a greenish tint of skin replaces whiteness.   The period of the growth of the body when great demands are made upon the functions of assimilation and nutrition is a favouring time for the occurrence of anæmia; and the debility and paleness of young people who "grow too fast" find here an explanation.

Increase in the quantity of blood may be considered here only in its local relations from overfilled blood-vessels, or what is called congestion of parts.   It is seen in the cheeks in the "hectic flush" of fevers, and in the early stage of an inflamed part, redness being one of the classical signs of inflammation.   The cause of the red rash in scarlet fever is congestion of the vessels of the skin, and its ready paling on pressure of the skin shows that the blood can be displaced from the vessels of the surface.   Permanent congestion, leading to permanent redness of parts, are not at all uncommon ; and the enlarged congested vessels can be traced on many a weatherbeaten face, or in the bloated, reddened features of the habitual drinker.   Temporary flushing of the cheeks—apart from blushing, but doubtless conveyed through the medium of the nervous system—is seen frequently in indigestion, and is the usual result of the ingestion of stimulants, as alcohol ; and an aggravated degree of this produces what is known as "nettle-rash."

When the blood cannot be perfectly oxygenated in the lungs, or when from any cause the return of blood to the heart is hindered, then the surface assumes more or less of a bluish or livid tint.   This is seen best in the parts which are most vascular, e.g., the mucous membranes, as the lips, or the regions most remote from the centre of the circulation, as the tips of the fingers, or toes, ears or nose.   Simple exposure to cold will do as much, causing temporary stagnation of the blood in these remote regions of the body.   Prolonged chilling of the surface is always followed by reaction ; the blood flows back to the part with increased vigour, and redness replaces pallor, and not only the redness of mere congestion of vessels, but that denoting inflammation; and the tender, swollen, reddened "chilblain" is thus related to the graver "frost-bite," where the effect of cold is seen in its severest form.

Jaundice is due to the pigment of the bile finding entrance into the blood, either because too much bile is manufactured by the liver than can be got rid of, or else, and more frequently, because the escape of the bile into the intestines is hindered. This colouring matter then stains all the tissues yellow, and the skin and the white part of the eye show the colour most plainly. If the cause cannot be removed, then the colour changes gradually from a deep orange-yellow to a more or less greenish or olive-green tint. But in such cases the cause is probably beyond removal. Young infants frequently, in the first few days after birth, before their respiratory and cutaneous functions are fully established, pass through a slight attack of jaundice. This seldom has any grave significance.

Passing to other forms of change of colour, we may note the curious, muddy or tawny, complexion of people who have suffered from ague or marsh fevers, a complexion which goes with them through life, and is probably dependent on the change which the spleen—an organ intimately connected with blood formation—undergoes in such diseases.

Lastly, certain substances when taken into the blood

will stain the skin and tissues. Of these, silver is the most notable example; and people who have taken this metal medicinally for a long period, are liable to have the skin of their face changed to a bluish leaden hue from the action of the light upon the silver. And one of the most marked effects of chronic lead-poisoning is the appearance of a line of bluish discoloration along the thin part of the gums where these border on the teeth.

Then many disorders of the skin which give rise to irritation, congestion, or inflammation, leave behind them permanent yellow or brownish marks of discoloration. There is one common affection which is accompanied by a yellowish pigmentation. It is due to the presence of a fungus on the skin, and occurs on the body favoured by clothing and uncleanliness. It is readily removable.

CHAPTER IX.

ON 'TEMPERAMENT,' 'HABIT,' AND 'TONE.'

FROM the earliest times it has been matter of faith among writers on medicine that all mankind can be subdivided into a few groups, each with sharply-marked characteristics, and each with certain tendencies to particular affections. How much truth underlies this doctrine it is needless to inquire. Let it suffice that even now we recognise the types of constitution which so long ago were first pointed out. Like all generalizations, however, these types or temperaments require to be dealt with liberally in drawing conclusions ; and in the majority of cases, an approximation to the characters can alone be found, and when found are not of much practical value. They are four in number, and the terms denoting them have crept into current use as descriptive of mental as well as bodily qualities, viz., sanguine, lymphatic or phlegmatic, bilious or melancholic, and nervous.

The *sanguine* is characterised by healthy vigour, activity of body, and of mind; with ruddy cheek, well developed muscles which are used with delight. There is a great deal of tone about the tissues, which are well braced up, and the bright aspect of the face is representative of the hearty condition of the mind, and the cheerfulness with which life is regarded. Reverses do not sadden, nor sorrow damp the brightness of the walk in life. This and more might be said of this happy, healthy condition, which is happy because healthy. The *lymphatic* person, however, is less carried above his life-conditions. He is pale, and generally fair in feature ; slow, and rather dull in mind. The *bilious* is typified by dark hair, a thick, dull skin, a sluggish mind, rather easily-provoked temper, and a liability to disorders of digestion which cloud the mind as well as affect the body. The *nervous* on the other hand is easily roused, constantly

occupied, but seldom continuously—ever seeking fresh
fields to work in, without in any of them pursuing his
work to the end. Highly intelligent, often precociously
so in youth, the mind is not well-balanced, so that the
individual is excitable in temperament, and both unduly
depressed by failure or unduly exalted by success.

Another term, also handed down to us from antiquity,
and far too widely employed, is "habit," or its correlative,
"constitution." For example, we talk of people being of
a "gouty constitution," or a "rheumatic habit," or as
being of a "strumous type," or "consumptive tendency,"
judging in these and many other instances from appearances, but meaning no more by the phrases than that the
individuals in question are disposed to such and such
particular class of disorders. In a broad and general
sense no doubt—as in the above-quoted instances—true
conclusions may be arrived at simply from such observation, for the affections they denote are generally dependent upon some inherited constitutional defects, and
are stamped upon the individual. But too rigid re-
liance upon certain prominent characters as indicative
of such tendencies is liable to lead to much erroneous
inference, and it is noteworthy that even the same
"habit" may be presented under wholly different characters. For example, the "gouty" man is quite as
often pale, thin, and flabby-textured, as plethoric and

heavy in build; and the strumous child as frequently fair-haired, oval-faced, and thin-skinned, as coarse-featured, and thick-lipped.* When, again, the term "habit" is applied to such diseases as "apoplexy," the error is very great; for it is notorious that the prevalent idea that individuals with thick, short necks and sanguine tempera-ment are the more liable to apoplexy is wholly unfounded in fact, that disease depending on causes equally at work among the lean and long-necked as among the former. Far better to erase the word "habit" from the vocabulary than run the risk of such misapplications.

We have more than once spoken of loss of "tone," and vague as the expression is, it is well to dwell on this as indicating perhaps better than anything else the state or condition which of all others is to be found on the side of unhealthiness. Everyone knows how much dis-comfort and inconvenience to themselves and to others is experienced by those who, suffering from no definite derangement of any organ, are yet in a state of ill health during the greater part of their lives. Unable to resist even the slightest influences, and feeling "well" only at rare intervals, there is something in their constitution which baffles definition. Their aspect is generally more or less pallid; for they take but little exercise, and

* It is usual to regard the strumous state as comprising these two distinct *types*.

7

undergo but few exertions, so that the circulation is slow, and cold extremities with consequent want of sleep are their bane. The skin is flabby, often moist, and the dark shadows under the eyes denote that the relaxed condition of the skin is sufficient to obscure its natural colour. The same flabbiness belongs also to their muscles; all exertion is painful and wearisome, as well as irksome, and a general debility and languor accompanies them. How many does not this typify, and how often is it largely the result of want of personal effort, and personal hygiene? Contrast such habits of body with those of the sturdy and vigorous, and it will be seen that in one the tonicity is deficient, in the other marked. Well-regulated, but not over-violent, open-air exercise, careful diet, frequent cold bathing, and above all some definite pursuit to engage the mind, will do more to restore the balance in the former than a course of medicinal treatment however tonic and however prolonged.

Our highly-artificial lives have much to do with the prevalent valetudinarianism of the present day; and the customs of society which run counter in so many ways to the teachings of common sense, have done more evil in this respect than can well be estimated.

THE END.   S. C.